YOUR KNOWLEDGE HAS VALUE

- We will publish your bachelor's and master's thesis, essays and papers

- Your own eBook and book - sold worldwide in all relevant shops

- Earn money with each sale

Upload your text at www.GRIN.com and publish for free

Bibliographic information published by the German National Library:

The German National Library lists this publication in the National Bibliography; detailed bibliographic data are available on the Internet at http://dnb.dnb.de .

This book is copyright material and must not be copied, reproduced, transferred, distributed, leased, licensed or publicly performed or used in any way except as specifically permitted in writing by the publishers, as allowed under the terms and conditions under which it was purchased or as strictly permitted by applicable copyright law. Any unauthorized distribution or use of this text may be a direct infringement of the author s and publisher s rights and those responsible may be liable in law accordingly.

Imprint:

Copyright © 2018 GRIN Verlag
Print and binding: Books on Demand GmbH, Norderstedt Germany
ISBN: 9783668676688

This book at GRIN:

https://www.grin.com/document/418785

Mutinda Jackson

Medicinal Drugs and Organic Chemistry

GRIN Verlag

GRIN - Your knowledge has value

Since its foundation in 1998, GRIN has specialized in publishing academic texts by students, college teachers and other academics as e-book and printed book. The website www.grin.com is an ideal platform for presenting term papers, final papers, scientific essays, dissertations and specialist books.

Visit us on the internet:

http://www.grin.com/

http://www.facebook.com/grincom

http://www.twitter.com/grin_com

AUTHOR: Jackson Mutinda

Medicinal Drugs and Organic Chemistry Relationship

Medicinal chemistry refers to the science dealing with the discovery along with design of novel therapeutic chemicals/bio-chemicals, and their respective development into useful medicines. It is a phenomenon that may comprise of compounds isolation from nature or new molecules synthesis, investigation of the connections between natural and/or synthetic compounds and their biological processes. Again, it may involve elucidations of these compounds' interactions with varied receptors such as DNA and enzymes, determination of their absorption, distribution and transport properties, and metabolic transformation studies of these chemicals, excretion and their toxicity.

Notably, since the 1960s, current techniques for discovering new drugs have incalculably evolved in line with phenomenal organic chemistry, physical chemistry, analytical chemistry, pharmacology, biochemistry and therapeutic advances. With respect to organic chemistry, it is evident that the modern tools applied in conducting drug design, substantially encompasses the entire vital processes of discovering drugs. Such courses include the application of physical principles, understanding the reaction of the body to the drug and the perceived synthetic organic processes used in the new compound preparation. However, drugs are not usually discovered, but rather a lead compound, which is a prototype compound with significant desirable features like those of pharmacological activities, though undesirable facets may also be present such as toxicity. Significantly, numerous studies have demonstrated a substantial relationship between organic chemistry and drugs as will be explored in this essay.

Apparently, even before looking at the modern drug discoveries, scholars have demonstrated a wide range of drugs that had happenstance discovery phenomenon other than the rational drug design. In its crudest sense, medicinal chemistry has been practiced in different regions for many centuries as man depended on herbs, plant roots and barks for disease treatments. For instance, one of the earliest medicines, the root *Dichroa febrifuga* (prescribed for fevers) was discovered by the Chinese Emperor-Sheng Nung around 5,100 years ago (Silverman, & Holladay, 2014). According to research, this plant has alkaloids, which are currently used in malaria treatment. Ma Huang, which is presently known as *Ephedra sinica* was earlier used as a heart stimulant, a diaphoretic agent and was recommended for asthma, hay fever alongside nasal and chest congestion treatment (An *et al.* 2014). Nowadays, this plant is renowned to have ephedrine (used as a stimulant, decongestant or appetite suppressant and hypertensive agent) and pseudoephedrine, used as a sinus or nasal decongestant and stimulant (pseudoephedrine hydrochloride shown in Figure 1.

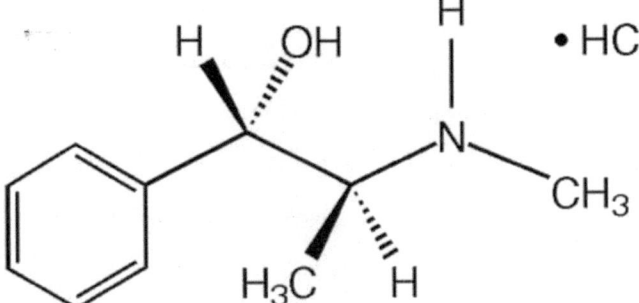

Figure.1-Pseudophedrine hydrochloride (Silverman, & Holladay, 2014)

Remarkably, one of the greatest early herbal medicines was discovered by a monk called Calancha in 1633, where the South-American Indians extracted the cinchona bark for chills and fevers (Rates, 2001). Later, the active constituent was isolated and determined to be quinine (shown in Figure 2), which is widely used as an antimalarial drug, not to mention that it also contains antipyretic and analgesic properties (Schreiber, 2011).

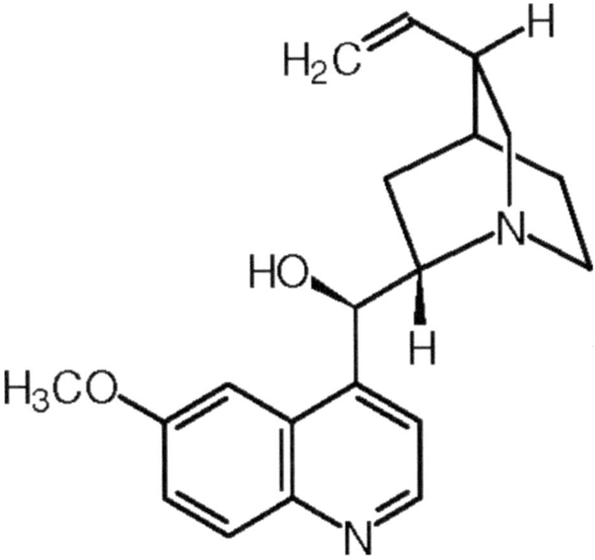

Figure 2-Quinine (Silverman, & Holladay, 2014)

Research shows that the modern therapeutics began with a foxglove plant extract cited in 1250 by Welsh physicians, named in 1542 by Fuchsius and introduced for dropsy (currently known as edema) treatment by withering in 1785 (An *et al.* 2014). As noted in Figures 3 and 4 below, the active constituents include glycosides from the foxglove plant (*Digitalis purpurea*) and *Digitalis lanata*, that is, digitoxin (R=H) and digoxin (R=OH), in that order, which are presently crucial drugs for congestive heart failure treatment (Silverman, & Holladay, 2014). Even today, digitalis (referring to all the cardiac glycosides,) is still manufactured through foxglove extraction and that of the related plants.

Figure 3- Digitoxin (R=H) **Figure 4-** Digoxin (R=OH) (Silverman, & Holladay, 2014)

Alexander Fleming observed a green mold growing in a *Staphylococcus aureus*, and he realized that the bacteria were lysed where the two converged, a phenomenon that led to penicillin discovery (Rates, 2001). According to scientists at that time, there are specific conditions that are required to produce this phenomenon, yet they approve that penicillin does not lyse the bacteria, but rather prevents them from growing (Schreiber, 2011). Later, Fleming's experiment was reinvestigated by Sir Howard Florey with the intent of producing penicillin could be topically and systematically administered. The initial mold was *Penicillium notatum* that yielded low penicillin, leading to substitution with *Penicillium chrysogenum*. Although several organic chemists and scientists disputed over the actual penicillin structure for years, the correct structure was illuminated in 1944. However, since a number of penicillin analogs-R group varied, were early on isolated, only two of these analogs are used today: R=PhOCH2 (penicillin V) and R=PhCH2 (penicillin G) (Silverman, & Holladay, 2014).

Notably, the first benzodiazepine tranquilizer drug, Librium (7-chloro-2-(methylamino)-5-phenyl-3-H-1,4-benzodiazepine 4-oxide; chlordiazepoxide HCl as seen in Figure 6, was discovered serendipitously.

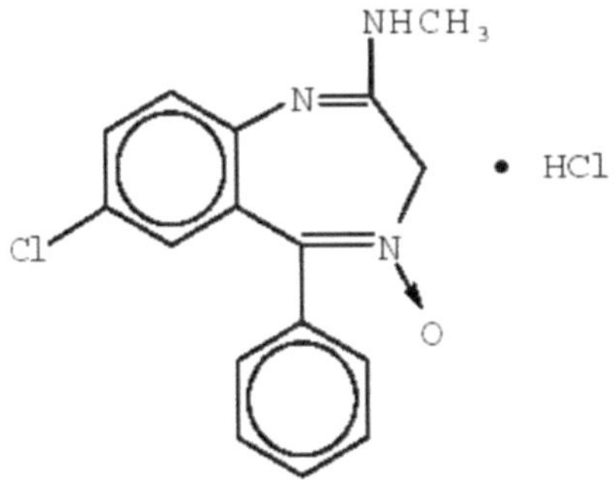

Figure 6- Chlordiazepoxide HCl (Silverman, & Holladay, 2014)

In this respect, Dr. Leo Sternbach was attempting to synthesize a novel group of tranquilizer drugs. Initially, Sternbach embarked on this program with the intention of preparing a benzheptoxdiazines, but when R^1 assumed CH_2NR^2 and R^2 remained C_6H_5, it was noted that the actual structure represented that of a quinazoline 3-oxide. Fortunately, further investigations that followed revealed that the compound was not a quinazoline 3-oxide, but instead it was the benzodiazepine 4-oxide, which was believed to have been produced in an unexpected corresponding reaction of chloromethyl quinazoline 3-oxide with methylamine (An et al. 2014).

Penicillin V together with Librium have been noted to be two crucial drugs that were discovered without a lead compound, although after their identification, they immediately became lead compounds for second-generation analogs. Nowadays, there are copious penicillin-derived antibacterials, which have been synthesized due the earliest penicillin structural elucidation (Silverman, & Holladay, 2014).

Diverse researches have also been performed to illuminate the idea that the character of the quaternary ammonium curare (conventional quaternary alkaloid poisons) can block the excitatory neurotransmitter acetylcholine action at muscle receptors. In recent analogous studies, it has been summed up that the physiological molecule action is a function of its chemical composition, with other scholars proceeding to suggest that the aliphatic alcohols hypnotic activity remains to be a function of their MW (Silverman, & Holladay, 2014). Thus, the two observations formed the future basis for medicinal chemists to focus on SARs. For example, the analysis as well as the description of SARs has been illustrated by the development of the sulfa drugs (sulfonamide antibacterial agents). Clinical trials determined that lead compounds exhibited diuretic alongside antidiabetic and antimicrobial activities, following preparation of numerous lead-compound sulfanilamide analogs (Schreiber, 2011).

To sum up, organic chemistry remains to be an august field, optimistically redefining itself in a manner that will hasten future therapeutics discovery. Over the years, biologists and chemists have been exploring the functions and structures of molecules in living systems, recently, emphasizing on the discovery of amino acids, glucose, neurotransmitters and even hormones. The knowledge of the forms of processes and targets that may be modulated, especially with small molecules will continue aiding the description of the principles underlying the rational therapeutics discovery. Succinctly, the continued advancement, redirection along with the evolution of this field plays an imperative role in the realization of human biology promise in the twenty-first century by scientists and chemists.

References

An, Y., Zheng, D. & Wu, J. (2014). An unexpected copper(II)-catalyzed three-component reaction of quinazoline 3-oxide, alkylidenecyclopropane, and water: *Chemical Communications, 50(1)*, 9165-9167.

Rates, S. (2001). Plants as source of drugs: *Toxicon, 39(2001)*, 603-613.

Schreiber, S. L. (2011). Organic synthesis toward small-molecule probes and drugs: *Proceedings of the National Academy of Sciences of the United States of America, 108(17)*, 6699–6702. doi:10.1073/pnas.1103205108

Silverman, R. B. & Holladay, M. (2014). *The Organic Chemistry of Drug Design and Drug Action*. Waltham: Academic Press.

YOUR KNOWLEDGE HAS VALUE

- We will publish your bachelor's and master's thesis, essays and papers

- Your own eBook and book - sold worldwide in all relevant shops

- Earn money with each sale

Upload your text at www.GRIN.com and publish for free